des&desi

Chemistry In My DREAMS

COLORING BOOK 1:
NATURAL PROCESSES

WRITTEN BY TERRESSA BOYKIN
illustrated by Billi French

ISBN: 978-1-9457-5102-8

Boykin, Terressa
des&desi
Chemistry In My Dreams

TABLE OF CONTENTS

PROLOGUE

In my dreams

Each night when I go to bed
To a dreamland in my slumber,

**With drooping eyes I slip
Into the learning land of wonder.**

In my dreams I see a wonderland
Of how and what things do,

And how we learned our chemistry
In elementary school.

THE LAZY BEE

The introduction to the honey-making process.

There was a comb of busy bees,
Each day they'd leave their hive,
And fly to find the nectar
From each flower, they would strive.

But there was a little lazy bee
Who would not work as the rest.
She stayed behind and slept all day
And never did her best.

While the lazy bee lagged behind,
Some busy bees were collectors,
And others guarded their hive,
Working as the protectors.

At the end of each day the busy bees
Brought nectar to their home.
They placed it in each empty cell
To make honey for their comb.

Each bee quickly flapped her wings
Fanning water from her nest,
Until the wet nectar had become
A sticky golden crest.

With the wetness in the nest now gone
The sweet nectar became gummy,
And dried into a sticky goo
A sweet and yummy honey.

But when the lazy bee came strolling home
Before the setting sun,
Little honey could she make,
No, not much and nearly none.

As the little bee thought and pondered
About what she did that day,
She remembered how she spent it,
Sleeping, relaxing and wasting it away.

Now the lazy bee understood
why so little she had earned.
"You should work, then play!" she said,
"Yes, bee busy!" She had learned.

So do your best in school each day
In reading, math and history.
You never know what you'll create,
Perhaps the next discovery.

THE SUNBURN SNOWMAN

The introduction to the water cycle.

I like the chilly temperature,
It helps me stay awake.
A roll of snow builds more of me
With each and every flake.

The brightness of the sunshine shows
My white and shiny mound,
But too much of the sunny rays,
Causes me to lose a pound.

Yes, I shrink in warmer weather
When the temperature is hot,
And if it's hot much too long,
I become a water spot.

Then as the sun rays shine on me,
My water puddle shrinks,
Until I turn into a gas
Almost invisible you'd think.

Each part of me slowly rises
Into the sky where I now meet,
The other water droplets from
Each snowman on our street.

We are now a water vapor
Rising up in a cloud we go,
Until the cloud is plump and full
With water vapor it starts to grow.

Soon our heavy weight
Makes us fall back to the ground below,
First as water droplets
And then as flakes of snow.

We have changed from solid to liquid
And then a liquid to a gas,
Until we drift down as new snowflakes
To play again with you at last.

EPILOGUE
In my dreams

Snuggling with my pillow
In this enchanted place,

Quietly, I am sleeping
With a smile upon my face.

In my dreams I see a wonderland
Of how and what things do,
All we learned of chemistry
In elementary school.

The night is almost over —
To another day I'll wake.
Until then I'll keep dreaming,
More to see, new friends to make.

THE END.

ACKNOWLEDGEMENTS

ABOUT THE AUTHOR

Terressa Boykin holds a Masters of Science in Organic and Kinetic Chemistry, has worked in the field of Polymer Chemistry and Chemical Engineering for over 28 years. In addition to volunteering in her community as a local and regional science fair judge, she is a member of the American Chemical Society's Coaching Program, the Education and STEM Committee for her local Chamber of Commerce and is the founder and CEO of Des&Desi Chemistry Classes for young children and the author of the children's book series Des&Desi *Chemistry In My Dreams* and *The Atom*. www.desndesi.com

ABOUT THE ILLUSTRATOR:

Billi French has a Bachelors of Arts Degree in Studio Art and a Certification in Illustration and Animation. She is the owner of the illustration company, BF Studios. www.billifrench.com.

CONCEPTUAL ARTIST:

Donna Merchant holds a Bachelor of Fine Arts Degree, majoring in art education. She is a resident artist/art teacher at Red Art & Design Studio. www.redartanddesign.com

MY INSPIRATION:

Desmen A. Boykin is a middle school student who has a passion for science and a talent for creative drawing. He hopes to be an engineer one day.

Desiree M. Mosley is an LPN and currently an RN nursing student who has a gift and a passion for helping others.

Other books by the author

DES&DESI
CHEMISTRY IN MY DREAMS
BOOK 1: NATURAL PROCESSES

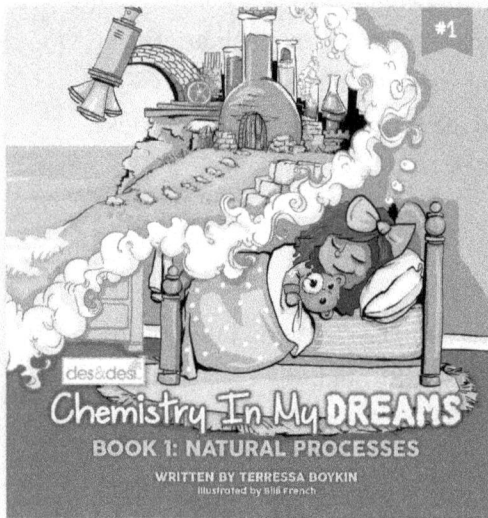

DES&DESI
CHEMISTRY IN MY DREAMS
BOOK 2: MAN-MADE PROCESSES

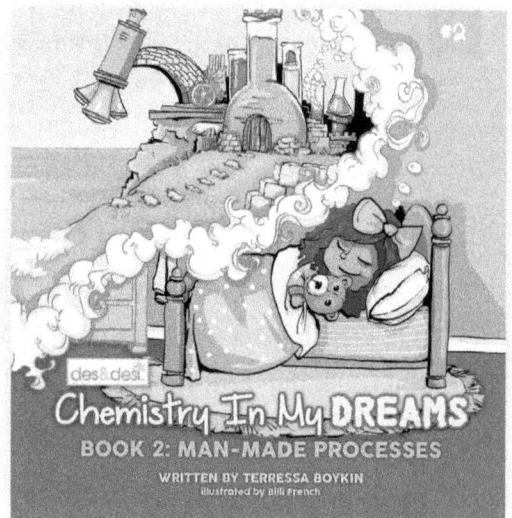

DES&DESI
MATTER:
IS ALL AROUND YOU

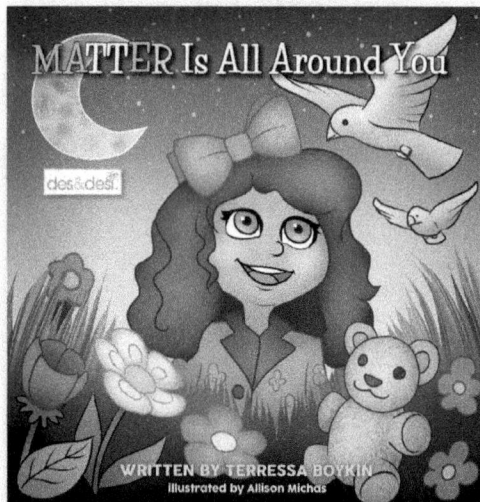

DES&DESI
THE ATOM
WHAT AM I REALLY?

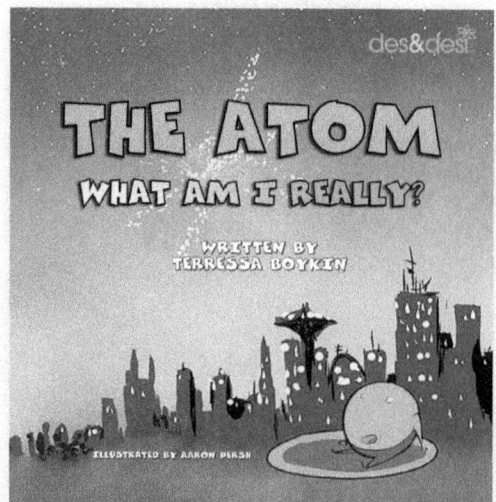